No-Nonsense Physics

Book 1

The Wave Basis
of
Special Relativity

Second Edition

Robert A. Close

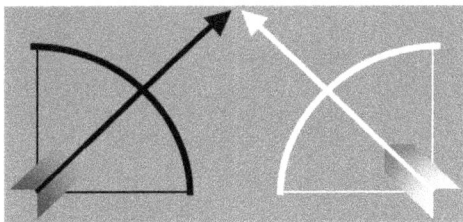

VERUM VERSA

Portland, Oregon

ISBN: 978-0-9837781-3-4

Published by Verum Versa (Portland, Oregon)

Contents

Contents

Preface

"An ocean traveler has even more vividly the impression that the ocean is made of waves than that it is made of water."
—Arthur S. Eddington

The theory of Special Relativity, as commonly understood today, was formulated by Albert Einstein and published in 1905 to explain the consequences of the experimental fact that the measured speed of light is independent of Earth's motion through space. In particular, it was clear that equations describing physical phenomena must have the property of "Lorentz covariance" in order to be valid for differently moving observers. Although the physical evidence requires only that the *equations* be Lorentz covariant, it has commonly been assumed that the underlying *space-time* on which the equations operate (i.e. the coordinate system) is also Lorentz covariant. Such a space-time is called "Minkowski space".

In this work we consider the simpler alternative that Lorentz covariance applies to the equations but not to the underlying space-time. This alternative assumption is more consistent with ordinary physical analysis. Physicists often derive Lorentz covariant equations from simple physical models in Galilean space-time (in which time and space are independent). Examples include transverse waves on a taut string and shear waves in an elastic solid. No serious scientist would claim that Lorentz covariance (with the appropriate wave speed) of the equation for displacements on a taut string implies that each string resides in its own separate universe with its own unique Minkowski space-time. Yet that would be equivalent to the logic of claiming that the Lorentz covariance of equations for matter implies that the universe is a Minkowski space.

Besides consistency with ordinary physics, the assumption of Lorentz covariant matter waves propagating in a Galilean space-time implies other properties of matter that have previously been considered as unrelated to

3

relativity. Classical waves are subject to an uncertainty relation exactly equivalent to the Heisenberg uncertainty principle. If production of matter and antimatter pairs is analogous to production of oppositely propagating waves in a solid, then matter and anti-matter must behave as mirror images of each other (and they do, although theorists like to think their relationship is more complicated than that).

Gravity has a simple interpretation if space-time is Galilean: it is simply the result of reduced wave speed in the vicinity of matter. This interpretation is consistent with general relativity, in which the change in light speed is directly related to the gravitational potential. Again, the issue is not the equations, but the interpretation of those equations.

In summary, the assumption of Lorentz covariant matter waves propagating in a Galilean space-time implies Special Relativity, the uncertainty principle, left-right asymmetry of matter and antimatter, and allows for a simple interpretation of gravity. By contrast, the assumption of Minkowski space-time is a "stand-alone" assumption: it does not imply Lorentz covariance of the equations for matter (that still requires appropriate physical processes), it does not imply left-right asymmetry of matter and anti-matter (a Nobel prize was awarded to the theorists who predicted that matter does not behave like its own mirror image), and the Minkowski space itself must be modified to accommodate gravity.

Hence there is good reason to regard the assumption of Galilean space-time as the simpler and more predictive one.

Additional comments for the Second Edition

In addition to the above rationale for a wave theory of light and matter, there is one more feature of waves that suggests a classical physics origin for matter and light. Waves have two types of momentum: intrinsic momentum of the medium that carries the wave, and "dynamical" or "wave" momentum associated with the transfer of energy via wave propagation. Until recently, little thought was given to the obvious corollary that waves must also have two types of angular momentum, just as elementary particles have both "spin" and "orbital" angular momentum. The classical analogues of "spin" and "orbital" angular momentum were first reported in 2009.[1] Spin density is related to velocity: it is the vector field whose curl is equal to twice the momentum density. Orbital angular momentum density is related to stress: it is the field whose time derivative is equal to minus the torque density. Saying that "the rate of change of spin angular momentum is equal to torque" is equivalent to saying that "the sum of spin and orbital angular momentum is constant."

Convergence between quantum and classical physics has also been progressing experimentally. Pilot waves, a theoretical concept invented in the 1920's as an interpretation of quantum mechanics,[2] have been produced in laboratory experiments by bouncing silicone droplets on a vibrating fluid surface.[3] These "walking droplets" exhibit many fundamental characteristics of quantum mechanical systems.[4]

There is still some way to go to achieve a complete understanding of the physics of elementary particles. My hope is that this book will help you on that journey.

— Robert Close (June 13, 2023)

---

[1] R. A. Close, "A Classical Dirac Equation," in *Ether Space-Time and Cosmology, vol. 3: Physical vacuum, relativity, and quantum physics.* M. C. Duffy and J. Levy, eds. (Apeiron, Montreal 2009) pp. 49-73.

[2] E. Madelung, Quantentheorie in Hydrodynamischer Form, Zeitschrift für Physik 40, 332 (1926). Louis de Broglie, "La nouvelle dynamique des quanta", in Solvay 1928, pp. 105–132.

[3] Yves Couder and Emmanuel Fort, "Single-particle diffraction and interference at a macroscopic scale," *Phys. Rev. Lett.*, **97**:154101, 2006.

[4] John W. M. Bush, "Pilot-wave hydrodynamics," *Annu. Rev. Fluid Mech.*, **47**:269–292, 2015.

I. Historical Introduction

"Ignorance is preferable to error; and he is less remote from the truth who believes nothing, than he who believes what is wrong." — Thomas Jefferson, *Notes on Virginia*

It is important to realize that scientists never formulate theories based solely on evidence. They are also influenced by prior knowledge and understanding (or misunderstanding). What follows here is a short summary of developments related to the theory of Special Relativity, based largely on works by Whittaker [1951, 1954].

Early attempts at a wave theory of light presumed that light waves propagate through a universal medium in the same manner as sound waves through air. This medium was dubbed the luminiferous 'aether'. Christian Huygens [1690] [Figure I–1] published an explanation of reflection and refraction based on the principle that each surface of a wave front can be regarded as a source of secondary waves.

Figure I–1: Christian Huygens (1629 - 1695)

Huygens also discovered that birefringent crystals can separate light rays into two distinct components, called polarizations. This effect is demonstrated in Figure I–2.

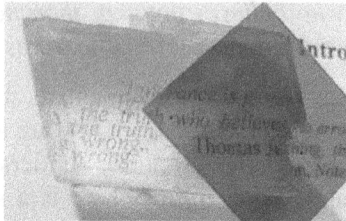

Figure I–2: Birefringent crystal. A polarizer can be rotated to select either of the two polarizations of light propagating through the crystal.

Isaac Newton [Figure I–3], among others, doubted that light was like sound waves because sound waves do not have this property of polarization. Nonetheless, Newton did perceive a similarity between color and the vibrations that produce sound tones. He also understood refraction of light rays in water well enough to explain the colors of the rainbow.

Figure I–3: Isaac Newton (1643 - 1727)

In 1675 Olaf Roemer attributed variations in the observed orbital periods of Jupiter's moons to variable light propagation distance between Jupiter and Earth. This interpretation, combined with Giovanni Domenico Cassini's parallax determination of interplanetary distances in 1672, determined the speed of light to be about 2.1×10^8 m/s (today the value is *defined* to be 2.99792458×10^8 m/s, so that the meter is actually determined by the speed of light rather than vice versa).

Because light, unlike particles, propagates at a characteristic speed, Thomas Young [Figure I–4] was convinced that light consists of waves. He demonstrated this wave nature by producing interference fringes from light passing through two narrow slits. Then in 1817 he explained polarization by proposing that light waves consist of transverse vibrations such as occur in elastic solids.

Figure I–4: Thomas Young (1773 - 1829)

Augustin Fresnel [Figure I–5] adopted Young's idea of transverse vibrations and developed a highly successful theory that explained diffraction and interference in addition to reflection and refraction. He supposed the aether to resist distortion in the same manner as an elastic solid whose density is proportional to the square of the refractive index.

Figure I–5: Augustin Fresnel (1788 - 1827)

A conceptual problem with a solid aether is the question of how ordinary matter can coexist and move freely through it. George Gabriel Stokes [Figure I–6] proposed that the aether was analogous to a highly viscous fluid or wax: elastic for rapid vibrations but fluid-like with respect to slow-moving matter. A modern-day equivalent would be "oobleck", a non-newtonian fluid made from corn starch and water. There are numerous entertaining videos of oobleck available online. The name "oobleck" is taken from the children's book *Bartholomew and the Oobleck*, by Dr. Suess. [1949]

Figure I–6: George Gabriel Stokes (1819 – 1903)

A more direct difficulty with the solid aether model was that density variations (e.g. at the interface between vacuum and medium) led to coupling between transverse and longitudinal waves, a phenomenon not observed for light waves. James MacCullagh [1839] [Figure I–7] avoided this problem by proposing a 'rotationally elastic' aether with inertial density ρ whose potential energy U depends only on the elastic modulus μ and the rotation angle (approximated by half the curl of displacement ξ):

$$U = \frac{1}{2}\mu(\nabla \times \xi)^2$$

The resulting wave equation is:

$$\rho\frac{\partial^2 \xi}{\partial t^2} = -\mu\nabla \times (\nabla \times \xi)$$

This is simply the equation of elastic shear waves. Matter was now presumed to alter the elasticity of the aether rather than its density. This model successfully accounted for all the known properties of light with $c^2 = \mu/\rho$.

Figure I–7: James MacCullagh (1809 – 1847)

Joseph Boussinesq [1868] [Figure I–8] later proposed that the aether could be regarded as an ordinary ideal elastic solid whose physical properties (density and elasticity) are unchanged by interaction with matter. The optical properties of matter were thus entirely due to the manner in which matter interacts with the aether. With this approach any classical optical phenomenon could be consistently modeled simply by finding the appropriate interaction term.

Figure I–8: Joseph Boussinesq (1842-1929)

Despite the successes of solid aether theories, scientists continued to pursue theories of a fluid aether through which solid matter could propagate. William Thomson (Lord Kelvin) [Figure I–9] attempted to model the aether as a 'vortex sponge': a fluid full of small-scale vortices with initially random orientation. He argued that this system could support transverse waves analogous to those in an elastic solid.

Figure I–9: William Thomson (Lord Kelvin, 1824 – 1907)

James Clerk Maxwell [Figure I–**10**] conceived of a different type of aether to derive the (Lorentz covariant) equations of electricity and magnetism. He modeled the aether as a network of rotating elastic cells interspersed with rolling spherical particles [Maxwell 1861a,b, 1862a,b].   The resultant equations for light waves are equivalent to those of MacCullagh.

Figure I–10: James Clerk Maxwell (1831 - 1879)

13

Maxwell's crude model is illustrated in Figure I–11. A flow of particles from **A** on the left to **B** on the right causes rotation of the elastic cells. This rotation is interpreted as a magnetic field pointing in opposite directions on each side of the current.

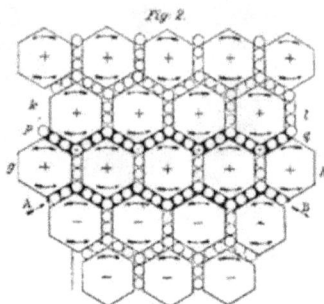

Figure I–11: Maxwell's aether model.

[Maxwell 1861b]

At this time, matter was still presumed to move through the aether as particles moving through a fluid. Many attempts were made to directly measure the relative motion between the earth and the aether. The most notable of these was an experiment first reported by Albert Michelson [Figure I–12] in 1881 and subsequently improved [Michelson and Morley 1887].

Figure I–12: Albert Michelson (1852 - 1931)

As Earth moves through a fluid aether, light propagating back and forth along a path aligned with the earth's motion should have a slightly slower average velocity than light propagating perpendicular to the earth's motion. Michelson and Morley attempted to detect the difference in propagation time for two light beams with a common source but propagating in perpendicular directions (with mirrors to bring the beams back together). The combined beams formed "interference fringes" that were expected to shift as their apparatus was rotated. However, no such effect was observed in this or other 'aether-drift' experiments.

Figure **I-13** shows how interference fringes are formed when waves are combined from two different sources.

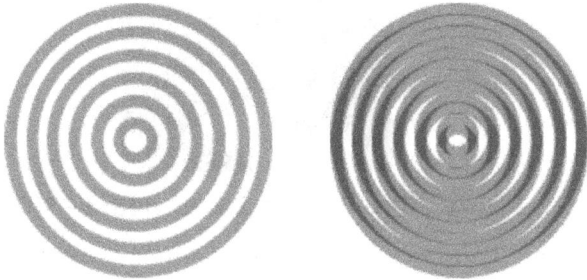

(a) Single wave source.   **(b) Two wave sources.**

Figure I-13: Waves from two different sources form "fringes". (a) Dark and light regions represent up and down motion of the wave, respectively. (b) With two sources separated vertically by ½ wavelength, waves propagate horizontally as if there were only a single source, but the waves cancel in the vertical directions.

For light waves shining on a screen, regions of wave oscillation are seen as bright spots, whereas regions without wave oscillation are dark.

Oliver Lodge [1893] demonstrated that the velocity of light is not noticeably affected by nearby moving matter, indicating that aether is not dragged along with matter. He also reported George FitzGerald's suggestion that the inability to measure motion relative to the aether could be explained if matter contracts along the direction of motion through the aether [Lodge 1892]. Joseph Larmor [1900] noted that in addition to the shortening of length, moving clocks should also run slower.

Hendrik Lorentz [1904] [Figure I–14] combined length contraction and time dilation to obtain the complete coordinate transformations.

Figure I–14: Hendrik Lorentz(1853 – 1928)

Henri Poincaré [1904] [Figure I–15] gave the name 'Principle of Relativity' to the doctrine that absolute motion is undetectable. He also deduced that inertia increases with velocity and that no velocity can exceed the speed of light.

Figure I–15: Jules Henri Poincare (1854 – 1912)

Albert Einstein [1905a] reformulated relativity with the more positive assertion that the speed of light is a universal constant independent of observer motion. Einstein's formulation of relativity is elegant because it yields physical results without reference to an invisible aether. Consistency with Einstein's "special theory of relativity" requires only that the equations for matter be Lorentz covariant (i.e. unchanged by Lorentz transformations).

Figure I–16: Albert Einstein (1879 - 1955)

In Einstein's theory of gravity, developed between 1911 and 1915, the speed of light is no longer regarded as strictly constant. Instead, the gravitational potential is a measure of the variation of the speed of light as seen from a fixed frame of reference (local measurements at different places still yield the constant speed c) [Einstein 1956 p. 93]. This implies that light waves refract in a gravitational field, a phenomenon first verified in 1919 when a group led by Arthur Eddington measured the shifted positions of stars whose light passed close to the sun during a solar eclipse [Dyson et al 1920].

One difficulty with the classical theory of light was a lack of success in describing radiation from a cavity at a fixed temperature (a 'black body'). Max Planck [1900] [Figure I–17] derived the correct formula for blackbody radiation by supposing light to be emitted by vibrators whose energy $E = nhf = n\hbar\omega$ is an integral multiple n of a constant h multiplied by the frequency f (or a multiple of $\hbar = h/2\pi$ times the angular frequency $\omega = 2\pi f$). Albert Einstein [1905b] used the idea that radiation consists of discrete quanta to explain the photo-electric effect, in which the frequency of light must exceed a certain threshold in order to liberate electrons from a metal.

Figure I–17: Max Planck (1858 – 1947)

Niels Bohr [1913] [Figure I–18] used quantization of angular momentum and energy to derive energy levels and spectral frequencies of the hydrogen atom.

Figure I–18: Neils Bohr (1885-1962)

Recognizing that quantization is often associated with waves and vibrations, Louis Victor de Broglie [1924] [Figure I–19] proposed in his doctoral thesis that electrons have a wave-like character with energy proportional to frequency $E = \hbar\omega$ and momentum proportional to wave vector $p = \hbar k$. Bohr's quantization of angular momentum is then equivalent to the requirement that stable electron orbits contain an integral number of electron wavelengths.

Figure I–19: Louis Victor de Broglie (1892-1987)

Soon afterward, Werner Heisenberg [1927] proposed an "uncertainty principle" for elementary particles. This uncertainty principle states that the statistical variances of momentum and position cannot have an arbitrarily small product.

Figure I–20: Werner Heisenberg (1902 - 1976)

If particles are regarded as waves, Heisenberg's uncertainty principle is simply a classical property of Fourier transforms. If a given number of wave cycles is decreased in spatial extent, then the wave number (inverse of wavelength) is increased. The uncertainty principle is obtained by applying this simple relationship to the spreads, or uncertainties, of both spatial position and wave number.

Walter Elsasser [1925] suggested that the wave nature of electrons would explain maxima and minima in the angular distribution of electrons scattered from a platinum plate in experiments reported by Clinton Davisson and Charles Kunsman. The wave nature of electrons was confirmed in 1927 when electron diffraction by crystals was clearly demonstrated in experiments by Davisson and Lester Germer [1927], and independently by George Thomson and Andrew Reid [1927].

Figure I–21 shows the wave interference fringes formed by diffraction of x-rays through a crystal. Bright regions of the image show where electron waves were large (many electrons), and dark regions show where the electron waves cancelled (few electrons).

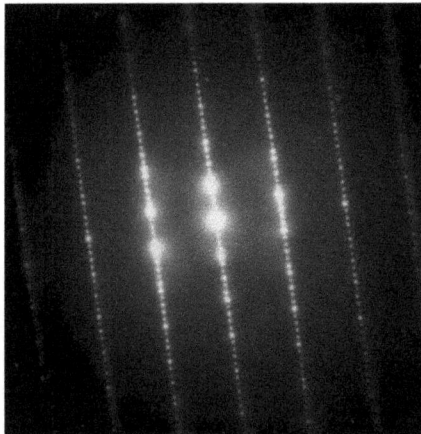

Figure I–21: Electron Diffraction Fringes

For a comparison of diffraction using x-rays and electrons, see:

http://online.cctt.org/physicslab/content/phyapb/lessonnotes/dualnature/Davisson_Germer.asp

Recent experimental work, starting in the Paris laboratory of Yves Couder, illustrates how waves can determine particle motion. These experiments utilize oil droplets bouncing on a vibrating pool of fluid to simulate wave-particle duality. A good summary of this work can be found at the website of MIT mathematician John Bush:

https://thales.mit.edu/bush/index.php/4801-2/

The discovery of the wave-like propagation of matter solves the historic dilemma of how matter can move freely through a solid aether. In addition, properties of the elastic medium itself need not change at all at the interface between vacuum and matter, thus explaining the lack of coupling to longitudinal waves.

The wave nature of matter also leads directly to the Principle of Relativity without any modification of the classical Galilean view of Euclidean space and absolute time, as will be shown below. However, mechanical modeling of fundamental physical processes was no longer in vogue when the wave nature of matter was discovered. Matter waves were not regarded as ordinary classical waves.

II. Measurements with Waves

"If we are to achieve results never before accomplished, we must employ methods never before attempted." — Francis Bacon

The first part of the following discussion closely follows Einstein's explanation of Special Relativity but with different rationale [Einstein 1956 pp. 29-37]. A point in spacetime is defined by a 3-dimensional position (x, y, z) and a time (t).

Let us consider the transformations between coordinates of relatively moving observers who measure distances by timing how long it takes for waves to propagate back and forth between two points.

The defining equation would be:

$$\left(\Delta\ell\right)^2 = \left(\Delta x\right)^2 + \left(\Delta y\right)^2 + \left(\Delta z\right)^2 = \sum_{i=1}^{3}\left(\Delta x_i\right)^2 = c^2\left(\Delta t\right)^2 \qquad (1)$$

where $\Delta\ell$ is the spatial distance between two points at a fixed time (the symbol "Δ" indicates "change of"), c is an arbitrary constant, and Δt is the time it would take to propagate a wave from one point to the other if they remained stationary. The time interval Δt could also be regarded as half of the time required for a wave to make a round trip between the two points. Simply put, the equation simply defines the constant "c" as the wave speed:

$$c = \frac{\Delta\ell}{\Delta t} \qquad (2)$$

By defining distance in terms of wave propagation time, the constant c is simply a scaling factor that relates the units of distance to the units of time. For now, let's suppose the waves are light waves.

The 'separation' s between two points in spacetime is defined by the equation:

$$s^2 = (c\Delta t)^2 - (\Delta x)^2 - (\Delta y)^2 - (\Delta z)^2$$
$$= (c\Delta t)^2 - (\Delta \ell)^2 \tag{3}$$

This is the difference between the squares of the spatial distance ($\Delta \ell$) between the points and the distance light would travel ($c\Delta t$) in the time interval between the two spacetime points. If $(\Delta \ell)^2 > (c\Delta t)^2$, then $s^2 < 0$ and the separation is said to be "spacelike". If $(c\Delta t)^2 > (\Delta \ell)^2$, then $s^2 > 0$ and the separation is said to be "timelike". Be aware that some authors define s^2 with the opposite sign.

Now consider propagation of a wave from point P_1 to point P_2. In a reference frame in which the points are stationary, Eq. 1 holds and $s^2 = 0$. This separation is said to be "lightlike". An observer in a different inertial reference frame using the same definition of distance would have, by definition:

$$\sum_{i=1}^{3} \left(\Delta x_i'\right)^2 = c^2 \left(\Delta t'\right)^2 \tag{4}$$

where the primes indicate the different coordinates.

For both observers, the separation is zero. Combined with an arbitrary offset, the invariance of this quantity for different observers is the condition that Lorentz used to derive the relativistic transformations. But the measured distances (or time intervals) would not be the same for the two differently moving observers. The reason is that the relative speed of a wave and a moving observer is different from the relative speed of a wave and a stationary observer. If the moving observer is unaware of their own motion, then measured distances and times will differ from those of a stationary observer.

II.1 Time dilation

For example, suppose a submarine navigator is using sonar both to measure time and to detect fish in the water as pictured in Figure II–1. The submariners use special sonar clocks that measure time by cycling sound wave pulses back and forth across a fixed distance in the water perpendicular to the direction of motion. Each cycle of wave transmission, reflection, and detection at the original site of transmission constitutes a tick of the clock. In this analysis we will neglect any effects of displacement of water by moving submarines. An animated presentation of this analysis may be found at:

http://www.classicalmatter.org/UnderwaterRelativity.htm

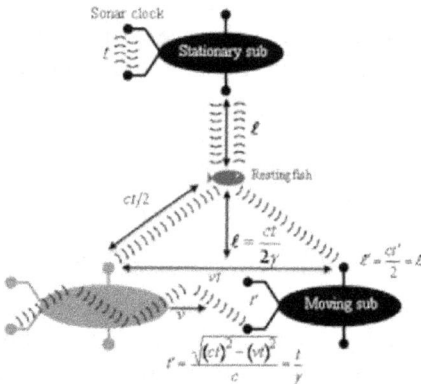

Figure II–1: Time Dilation: The clock on the moving sub O' ticks slower than the clock on the stationary sub O by the factor $1/\gamma = \left(\sqrt{c_s^2 - v^2}\right)/c_s$ because waves travel farther between transmission and detection. Both O and O' measure the same number of clock cycles for a wave to propagate from their own sub to the fish and back. Hence, they agree on distances perpendicular to the direction of relative motion. Here $c = c_s$ = sound speed.

First consider the submarine pictured at the top of Figure II–1 and the fish below it. If both the sub and the fish are at rest in the water, a sound wave reflected from the fish at distance ℓ would return after time $t = 2\ell/c_s$, where c_s is the sound speed. The distance to the fish is therefore taken to be $\ell = c_s t/2$. Suppose now that a second sub is moving in the water with speed v perpendicular to the original direction of wave propagation (pictured at the bottom of Figure II–1). The path of the sonar clock wave forms two sides of a triangle for each cycle. A similar triangle is formed by the wave propagation to the fish and back (the fish as viewed from the moving sub would appear to be moving backward). Therefore, the number of clock ticks which occur during wave propagation to the fish and back is independent of speed. This means that measurements of distance perpendicular to the direction of motion are unaffected by the motion.

If the second navigator doesn't realize that she is moving (and is not puzzled by fish swimming backwards), she would assume the same relation between distance and time: $\ell' = \ell = c_s t'/2$. But the navigator of the stationary sub would say that the moving sub's sonar wave propagated over a distance:

$$d \equiv c_s t = 2\sqrt{\ell^2 + \left(\frac{vt}{2}\right)^2} \qquad (5)$$

Substituting $\ell = c_s t'/2$ and solving for t' yields:

$$t' = t\sqrt{1 - v^2/c_s^2} = \frac{t}{\gamma} \qquad (6)$$

where:

$$\gamma = \left(1 - v^2/c_s^2\right)^{-1/2} \qquad (7)$$

Equation (6) merely expresses the fact that the clock on the moving submarine ticks more slowly than the stationary clock because the waves have farther to travel between ticks. Hence the time (t) measured by the stationary observer is longer than the time (t') measured by the moving observer. This phenomenon is referred to as "time dilation".

It is obvious that if the unprimed observer is truly stationary with respect to the water, then the moving clock does in fact tick more slowly. This is not merely an illusion. What is interesting is that the wave measurements performed by these submarines are insufficient to determine which sub is actually moving with respect to the water. The moving sub would interpret the stationary sub's clock as running slowly, and in this case the effect is an illusion. This point will be discussed later in connection with Doppler shifts.

Since the stationary navigator sees the fish (and first sub) move a distance $x = vt$ while the wave is propagating, the above equation can be rewritten as:

$$t' = \frac{t\left(1 - v^2/c_s^2\right)}{\sqrt{1 - v^2/c_s^2}} = \frac{t - vx/c_s^2}{\sqrt{1 - v^2/c_s^2}} \tag{8}$$

This is the Lorentz transformation of time between two observers, with the primed observer moving in the x-direction with velocity $+v$ with respect to the unprimed observer.

Since both observers measure the same distance to the fish ($\ell' = \ell$) the transformation of coordinates perpendicular to the motion must be simply:

$$y' = y$$
$$z' = z \tag{9}$$

II.2 Length contraction

Now suppose that one sub and fish are moving relative to the other sub parallel to the direction of wave propagation [Figure II–2].

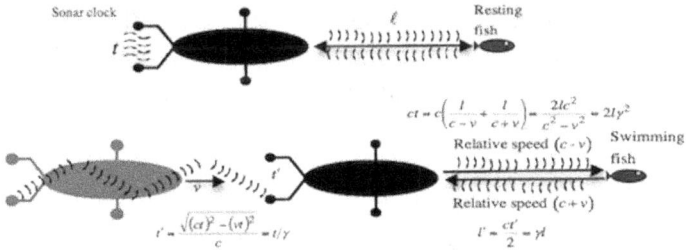

Figure II–2: Length Contraction: The true wave propagation time for the co-moving sub and fish is longer than for the stationary sub and fish by the factor $1/(1 - v^2/c^2)$**. Since the moving clock runs slow, the perceived propagation time is longer only by the factor** $1/\sqrt{1 - v^2/c^2}$ **. Hence the stationary sub observes a shorter length than the moving sub. Here** $c = c_s =$ **sound speed.**

As seen by the stationary sub, the angular frequency ($\omega = 2\pi/T$ where T is the period for one cycle) of the sonar clock on the first sub is slow according to Eq. 6 since the measured time t' is proportional to the moving clock frequency ω' times the absolute time t:

$$\omega' = \omega\sqrt{1 - v^2/c_s^2} \tag{10}$$

Another way to understand this is that everyone must agree on the phase of the clock at any instant: $\omega't' = \omega t$.

The absolute distance between the fish and sub remains constant at ℓ. However, the relative speed between the outgoing wave and the target fish is $(c_s\text{-}v)$ whereas the relative speed between the sub and the incoming wave is (c_s+v). Therefore, the absolute propagation time is:

$$t = \frac{\ell}{(c_s + v)} + \frac{\ell}{(c_s - v)} = \frac{2\ell}{c_s (1 - v^2 / c_s^2)} \qquad (11)$$

Of course, the moving sub still uses the relation $\ell' = c_s t'/2$. Substituting the temporal relation $t'/t = \sqrt{1 - v^2/c_s^2}$ yields the relation between lengths:

$$\ell' = \frac{c_s t \sqrt{1 - v^2 / c_s^2}}{2} = \frac{\ell}{\sqrt{1 - v^2 / c_s^2}} = \gamma \ell \qquad (12)$$

The stationary observer measures a shorter length (ℓ) than the moving observer (ℓ'). This phenomenon is known as "length contraction". In this case the moving observer measurement is artificially long due to the fact that the actual sound velocity relative to the observer is not the same for the outgoing and incoming directions. Since the wave propagates for a longer time in the direction of slower relative motion, the effect is an apparent increase in length relative to a stationary observer. Again, however, it is important to realize that the wave measurements alone do not determine which observer is moving (due to Doppler shifts).

As noted previously, the origin of the moving frame corresponds to $x = vt$ in the stationary frame. Therefore, the coordinate transformation is obtained by $\ell' \to x'$ and $\ell \to x - vt$:

$$x' = \frac{x - vt}{\sqrt{1 - v^2 / c_s^2}} \qquad (13)$$

This is the Lorentz transformation of position along the direction of motion.

II.3 Lorentz transformations

The Lorentz transformations derived above for time and distance have the same form no matter what type of wave is used for measurements. From here on we will simply denote wave speed by "c" no matter what type of wave is used. The type of wave under consideration should be obvious from context.

It is customary to use the definitions:

$$\beta = v/c$$
$$\gamma = (1 - \beta^2)^{-1/2} \tag{14}$$

A useful identity is:

$$\gamma^2 = \left(1 - \beta^2\right)^{-1} = 1 + \beta^2 \gamma^2 \tag{15}$$

Using the above expressions, the <u>Lorentz transformations</u> become:

$$ct' = \gamma ct - \beta \gamma x$$
$$x' = \gamma x - \beta \gamma ct$$
$$y' = y$$
$$z' = z \tag{16}$$

The inverse transformations merely change the sign of v (or β):

$$ct = \gamma ct' + \beta \gamma x'$$
$$x = \gamma x' + \beta \gamma ct'$$
$$y = y'$$
$$z = z' \tag{17}$$

These transformations were derived for sonar waves, but are valid for any wave measurements, including light waves. They are also valid for measurements made with material clocks and rulers, suggesting that material objects are closely related to light waves. This relationship will be explored later.

II.4 Four-vectors

Quantities that transform according to the Lorentz transformations are called 'four-vectors'. Each four-vector has three spatial components and a temporal component. The coordinate four-vector can be expressed as either $x^\alpha = (ct, x, y, z)$ or $x_\alpha = (ct, -x, -y, -z)$. This notation allows the separation to take on the beautiful form:

$$s^2 = x^\alpha x_\alpha$$

Other examples of four-vectors (with respect to light waves) include:

$$(\gamma c, \gamma \mathbf{v}) \qquad \text{Four – velocity}$$

$$(E/c = \gamma m_0 c, \mathbf{p} = \gamma m_0 \mathbf{v}) \qquad \text{Energy, momentum}$$

$$(\rho c, \mathbf{J}) \qquad \text{Electromagnetic charge, current}$$

II.5 Lorentz covariance

Now suppose that all the waves we are interested in satisfy an equation for a function $f(x, y, z, t)$ of the form:

$$\left[\frac{\partial^2}{\partial t^2} - c^2 \left(\frac{\partial^2}{\partial x^2} + \frac{\partial^2}{\partial y^2} + \frac{\partial^2}{\partial z^2} \right) + M^2 \right] f = 0 \qquad (18)$$

Where c is the wave speed and M is a constant. This is called the Klein-Gordon equation. The ordinary wave equation is a special case with $M = 0$.

Applying the Lorentz transformations for relative velocity $\mathbf{v} = v_x \hat{\mathbf{x}}$, the chain rule for derivatives yields:

$$\frac{\partial^2}{\partial t^2} \rightarrow \left(\frac{\partial t'}{\partial t} \right)^2 \frac{\partial^2}{\partial t'^2} + \left(\frac{\partial x'}{\partial t} \right)^2 \frac{\partial^2}{\partial x'^2} = \gamma^2 \frac{\partial^2}{\partial t'^2} + (\beta \gamma c)^2 \frac{\partial^2}{\partial x'^2}$$

$$\frac{\partial^2}{\partial x^2} \rightarrow \left(\frac{\partial t'}{\partial x} \right)^2 \frac{\partial^2}{\partial t'^2} + \left(\frac{\partial x'}{\partial x} \right)^2 \frac{\partial^2}{\partial x'^2} = \left(\frac{\beta \gamma}{c} \right)^2 \frac{\partial^2}{\partial t'^2} + \gamma^2 \frac{\partial^2}{\partial x'^2}$$

Substitution into the Klein-Gordon equation with $f'(x', y', z', t') = f(x, y, z, t)$ yields:

$$\left[[\gamma^2 - (\beta\gamma)^2] \left(\frac{\partial^2}{\partial t'^2} - c^2 \frac{\partial^2}{\partial x'^2} \right) - c^2 \left(\frac{\partial^2}{\partial y'^2} + \frac{\partial^2}{\partial z'^2} \right) + M^2 \right] f' = 0$$

Since $\gamma^2 - (\beta\gamma)^2 = 1$, this is simply the same Klein-Gordon equation in the transformed variables. The Klein-Gordon equation is said to be <u>Lorentz covariant</u> because its Lorentz transformation has the same form as the original equation. Differently moving observers agree on the wave velocity if all their measurements are made using the same type of wave.

II.6 One-way light speed

Occasionally, someone will claim to have found a way to measure a "one-way" light speed corresponding to either $(c - v)$ or $(c + v)$. In other words, they claim to be able to determine an absolute velocity v in space. Such claims have not withstood scrutiny. The problem is that, according to the Lorentz transformations, time in one reference frame varies with position in another reference frame. In fact, events that one observer sees as simultaneous at two different places would apparently occur at different times according to a relatively moving observer. This phenomenon is called the '<u>relativity of simultaneity</u>'.

Suppose a wave pulse is emitted from the moving sub at time $t = t' = 0$ (primed variables are in the moving reference frame). Denote the sonar wave speed by c. A stationary observer sees the fish ahead of the sub by distance $\ell = \ell'/\gamma$. The wave approaches the fish with relative speed $c - v$, so the wave reaches the fish at time $t = \ell/(c - v)$. By this time, the fish has moved to position $x = \ell + vt$ according to a stationary observer.

Applying the Lorentz transformations derived above, the co-moving observer sees the pulse reach the fish at position x' and time t':

$$x' = \gamma\left(x - \frac{v}{c}ct\right) = \gamma(\ell + vt - vt) = \ell'$$

$$t' = \gamma\left(t - \frac{v}{c}\frac{x}{c}\right) = \gamma\left(\frac{\ell}{c-v} - \frac{v}{c^2}(\ell + vt)\right) \quad (19)$$

$$= \gamma\frac{\ell}{c} = \frac{\ell'}{c}$$

As expected, the co-moving observer sees the wave reach the fish in exactly half of the round-trip time. One-way wave measurements are entirely consistent with round-trip wave measurements.

The above analysis shows that the Lorentz transformations are consistent but doesn't fully explain the time measurements. Why can't the moving observer figure out that it takes longer for the wave to travel away than it takes to travel back? Let's try!

Start with two clocks synchronized at the moving sub, then move one clock ahead slowly with speed δv (as seen by a stationary observer) until it reaches the fish. Since it is moving faster, the advancing clock ticks slower, losing time $(\delta t')$ at the rate:

$$d(\delta t') = \frac{dt}{\gamma_{v+\delta v}} - \frac{dt}{\gamma_v} \approx dt\left(\frac{d}{dv}(1 - v^2/c^2)^{1/2}\right)\delta v$$

$$= dt\left(-\frac{\gamma_v v}{c^2}\right)\delta v \quad (20)$$

where the subscripts v and $v + \delta v$ indicate the velocity that enters the expression for the Lorentz factor γ. The approximation becomes exact as $\delta v \to 0$. We would not expect the clocks to be synchronized if δv were large, since then the slowing would be significant even if the clocks were in a stationary reference frame.

The absolute time it takes to reach the fish is $dt = \ell/\delta v$. Therefore, the time lost relative to the sub on the moving clock as it reaches the fish is:

$$\delta t' = -\gamma_v \beta \ell/c \quad (21)$$

This exactly corresponds to the offset in the Lorentz transformation for time with $x = \ell$ (see Eq. 16). Separating clocks in a moving reference frame causes them to lose synchronization (as seen by a stationary observer), thereby making it impossible to distinguish between one-way and two-way wave measurements.

II.7 Length and time standards

We have seen how Lorentz transformations can be obtained by using sonar or any other type of wave to measure time and distance. The fact that measurements are related by Lorentz transformations is not a property of time and space per se. Rather it results from the methods used to measure time and distance. If the above-mentioned sailors were to rendezvous to share their data and some vodka, they might conclude after a few drinks that absolute time and space in moving underwater reference frames are related by Lorentz transformations using the speed of sound in water. After sobering up, however, they would realize that sonar is not the only way to measure time and distance, and their measurements are not evidence of any non-classical properties of underwater space-time.

The sonar clock might seem like an odd sort of clock, but consider the standard definition of a second, which is 9,192,631,770 periods of the radiation corresponding to the transition between the two hyperfine levels of the ground state of the cesium 133 atom [Taylor 1995]. If we regard the cesium atom as a kind of optical cavity that resonates at the prescribed frequency, then this is quite similar to our sonar clock.

Consider also that the standard definition of the meter is the length of the path traveled by light in vacuum during a time interval of $1/c = 1/299,792,458$ of a second [Taylor 1995]. So, we do in fact equate length with wave propagation time just as our hypothetical submariners do, and the quantity c is nothing more than a unit conversion factor.

II.8 Doppler shifts

Thus far we have shown that when waves are used to measure distance and time, the space-time coordinates transform between relatively moving observers according to the Lorentz transformations. Transformation of other dynamical variables is straightforward.

Suppose we have a plane wave described by the expression:

$$s = s_0 \cos(\mathbf{k} \cdot \mathbf{x} - \omega t) \tag{22}$$

Where ω is angular frequency (ωt changes by 2π radians each cycle) and $\mathbf{k} = \hat{\mathbf{k}}(2\pi/\lambda)$ is the wave vector for wavelength λ and direction unit vector $\hat{\mathbf{k}}$. The phase $\varphi = \mathbf{k} \cdot \mathbf{x} - \omega t$ of the plane wave is independent of observer motion. Therefore, stationary and moving observers see phases related by:

$$\mathbf{k}' \cdot \mathbf{x}' - \omega' t' = \mathbf{k} \cdot \mathbf{x} - \omega t \tag{23}$$

For motion along the x-axis, we can plug in the inverse transformations for x and t to obtain:

$$k_x' x' - \omega' t' = k_x(\gamma x' + \beta \gamma c t') - \omega(\gamma t' + \beta \gamma x'/c)$$
$$k_y' y' = k_y y'$$
$$k_z' z' = k_z z' \tag{24}$$

The coefficients of t' must be equal on both sides of the equation, and likewise for the coefficients of x', y', and z'. Therefore:

$$\omega' = \gamma \omega - \beta \gamma c k_x$$
$$k_x' = \gamma k_x - \beta \gamma \, \omega/c$$
$$k_y' = k_y$$
$$k_z' = k_z \tag{25}$$

Defining the vector $\boldsymbol{\beta} = \mathbf{v}/c$, the transformation for arbitrary direction of relative velocity is:

$$\omega' = \gamma(\omega - \boldsymbol{\beta} \cdot c\mathbf{k})$$
$$ck'_{||} = \gamma\left(ck_{||} - \beta\omega\right) \qquad (26)$$
$$\mathbf{k}'_{\perp} = \mathbf{k}_{\perp}$$

Hence the spatio-temporal frequency components $(\omega, c\mathbf{k})$ transform in the same manner as the coordinates $(ct, \mathbf{x})$ and therefore form a four-vector.

For simple plane waves $|c\mathbf{k}| = \omega$. Hence the frequency and wave vector transformations for motion parallel to $\mathbf{k}$ can be written as:

$$\omega' = \gamma\omega\left(1 - \beta\right) = \omega\sqrt{\frac{1-\beta}{1+\beta}}$$
$$k'_{||} = \gamma k_{||}\left(1 - \beta\right) \qquad (27)$$
$$\mathbf{k}'_{\perp} = \mathbf{k}_{\perp} = 0$$

The first of these equations is the relativistic Doppler shift formula for light waves.

The relativistic Doppler shift has a simple interpretation. First, consider the classical Doppler shifts as shown in Figure II–3.

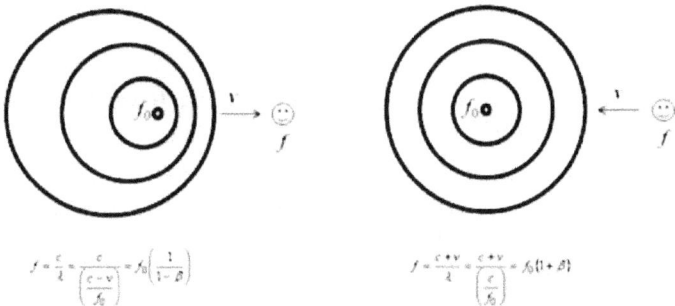

$$f = \frac{c}{\lambda} = \frac{c}{\left(\frac{c-v}{f_0}\right)} = f_0\left(\frac{1}{1-\beta}\right) \qquad\qquad f = \frac{c+v}{\lambda} = \frac{c+v}{\left(\frac{c}{f_0}\right)} = f_0(1+\beta)$$

Figure II–3: Classical Doppler shifts for moving (approaching) source and detector differ by a factor of $(1 + \beta)(1 - \beta) = 1/\gamma^2$. This factor is not affected by reversal of the velocity direction.

The Doppler shifts are different depending on whether the wave source or receiver is moving. If the source is moving and the receiver is stationary, then the speed of the waves is uniform, but the wavelength varies with direction. If the source is stationary and the receiver is moving, then the wavelength is constant, but the speed of the waves relative to the receiver varies with direction.

Consider a stationary observer O in a lighthouse which pulsates with angular frequency ω as in Fig. II-4.

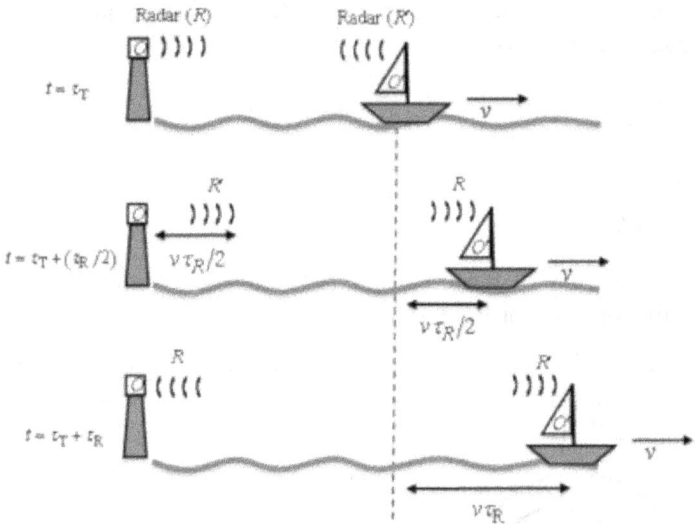

Figure II–4: Velocity Measurement: Radar signals sent simultaneously by O and O' will also be received simultaneously after reflection. Although O''s clock ticks slowly, the proportionality between radar pulse propagation time and total time elapsed is the same as for O. Therefore, both O and O' measure the same relative velocity.

An observer O' moves away from the lighthouse in a boat starting at $t = 0$. As a moving detector, O' receives a classically Doppler-shifted frequency of $\omega(1 - \beta)$. However, O' 's clock is running slow by the factor $1/\gamma$

because the boat is moving (and therefore subject to time dilation of Section II.1). Hence O' perceives the incident wave frequency to be higher by the factor γ so that $\omega' = \gamma\omega(1 - \beta)$. The stationary observer O would agree with this correct description of events. Note that observer O can measure the speed of observer O' by measuring the time of flight of radar pulses which reflect off of O ' and back to O. Successive pulses separated by transmission time interval τ_T will be received with delay time interval $\tau_R = \tau_T(1 + v/c)$ yielding $v = c\,(\tau_R - \tau_T)/\tau_T$.

Conversely, the observer O' incorrectly believes that the boat is stationary, and that O is moving. O' measures the speed of recession of the lighthouse via radar. The true propagation time of each pulse is the same as measured by O (see Figure II–4 above). The fact that O' 's clock is running slowly reduces measured times by the factor $1/\gamma$, but this does not affect the proportionality between the transmission time interval and the reception time interval. Therefore O' sees O recede with speed v.

Observer O' observes the lighthouse light fluctuate with angular frequency $\omega' = \gamma\omega(1 - \beta)$. This formula accounts for slowing of the moving clock and Doppler shift at the moving (receding) receiver. O' presumes the detected frequency to be classically Doppler shifted at the source by a factor of $1/(1 + \beta)$. Correcting for this Doppler shift yields $\omega'(1 + \beta)$ for the co-moving source frequency. Since O' thinks that O's clock is slow, the correction factor γ is again introduced to obtain the frequency perceived at the source. This leads to:

$$\omega = \gamma\omega'\left(1 + \beta\right) = \omega'\left(\sqrt{\frac{1+\beta}{1-\beta}}\right) \tag{28}$$

This is simply the inverse relativistic Doppler shift (inverse Lorentz transformation with $\omega' = ck'$). Note that O' incorrectly attributes the Doppler shift to a moving source rather than a moving detector, thereby resulting in an erroneous factor of $(1 + \beta)(1 - \beta) = 1/\gamma^2$. However, this mistake is exactly compensated by the fact that O' incorrectly believes that O's clock is running *slower* by the factor $1/\gamma$ when in fact it is running *faster* by the factor γ. O' mistakenly multiplies by γ instead of dividing by γ to correct for the different clock rates (an erroneous factor of γ^2). The erroneous factors of γ^2 and $1/\gamma^2$ cancel and O' correctly deduces the frequency ω for the stationary source at O. This cancellation of errors renders impossible the determination of motion relative to the medium that carries the wave. It is a rare case in which two wrongs make a right, and it is the crux of Special Relativity.

If the relative motion is not along the line of separation, then the Doppler shifts are dependent on angle as given in Eq. 26. Like the Lorentz transformations, relativistic Doppler shifts are a natural consequence of measurements with waves.

II.9 Acceleration

The above analysis demonstrates that wave measurements cannot determine absolute velocity. However, it *is* possible to determine absolute acceleration. Two observers undergoing a change in relative velocity can determine which of them is accelerating because only the accelerating observer will experience a force. If the inertial (constant velocity) observer sees that an accelerated object has changed its length and clock rate, that observer can reasonably conclude that the acceleration caused real changes to the object. Consistency therefore demands that the accelerated observer should attribute any observed changes in length and clock rate of distant objects to changes in his own accelerated rulers and clocks.

A special type of acceleration is rotation, for which every part of an object accelerates toward the axis of rotation. The force causing this acceleration can be experienced directly (e.g. stress caused by radial expansion of the rotating object) or by analyzing motion to identify the centripetal acceleration. For example, an ice skater doing a pirouette with arms outstretched can pull in their arms and observe that the rotation rate with respect to the surroundings increases. You can even do this without skates – try moving your arms in and out while spinning around on your feet.

The issue of absolute acceleration is not settled, however. Einstein argued in favor of a version of "Mach's principle" that an object's inertia is entirely determined by interaction with other masses in the universe. [Einstein 1956 p. 107] If true, this implies that rotation can only be determined relative to other masses in the universe, and not relative to some absolute space. In other words, the properties of space are entirely determined by the distribution of mass in the universe. Whether or not this view is correct is a topic in the realm of general, not special, relativity. We will not discuss it further here.

III. Matter Waves and Light

"It is better to light one small candle than to curse the darkness." — Confucius

The previous discussion of sonar waves has limited validity because sound waves in water are too simple to serve as a model of matter. One obvious problem is that the water surrounding a moving submarine is not stationary. We neglected the motion of the water because we eventually want to model matter as waves, and waves do not drag the medium with them.

A second problem is that the sonar clock had to be oriented perpendicular to the direction of motion so that its apparent length was independent of velocity. This limitation could be eliminated if the clock length varied relative to the direction of motion, so that the measured length (light time-of-flight) remains constant.

The third problem is that sound waves are scalar waves, described by a single number (e.g., pressure) at each point. A more interesting medium to consider is an elastic solid, which can support shear waves whose amplitude (displacement or rotation) can have multiple components. Recall that light waves were historically modeled as waves in an elastic solid.

The previous results show that the equations of Special Relativity are applicable to a wide variety of wave phenomena. The Lorentz transformations relate wave measurements made in different frames of reference. It is well known (and easily verified) that any wave equation of the form:

$$\left[\frac{\partial^2}{\partial t^2} - c^2 \nabla^2 + M^2 \right] f = 0 \qquad (29)$$

with constant scalar M is invariant in form under Lorentz transformations. In other words, Lorentz covariance is a general property of waves and not specific to light or other electromagnetic waves (except for the numerical value of the wave speed c).

Now we are in a position to appreciate what is special about light. Ordinarily we do not measure distances and times by propagating waves back and forth. Instead, we use material clocks and rulers. The amazing thing about material clocks and rulers is that the resulting distance and time measurements transform with exactly the same Lorentz transformations as would be obtained if the measurements had been made by propagating light waves. In other words, matter behaves as if it consists of waves that propagate at the speed of light. Since matter can appear to be stationary, we may suppose that the waves are either standing waves or that they somehow propagate in cyclic paths in the 'rest' frame. For example, the scalar wave equation has spherical harmonic solutions with dependence on azimuthal angle $(\phi - \phi_0)$ proportional to $\cos(\omega t \pm m(\phi - \phi_0))$, which does describe waves propagating in circles (azimuthal phase velocity increases with radius). Mass has previously been interpreted as circular motion at the speed of light [Hestenes 1990].

Historically, the equations of relativity were derived from the observation that absolute motion is indeterminable. Einstein formulated Special Relativity on the basis that the speed of light is independent of observer motion. Yet now we have a simpler alternative postulate for Special Relativity: matter consists of waves that propagate at the speed of light. Mathematically, this postulate is that equations for matter and light are Lorentz covariant.

This physical picture suggests that matter and anti-matter can annihilate into photons and vice versa because photons and matter are simply different packets of the same type of wave. The Heisenberg uncertainty principle, which seems so strange for particles, is simply the classical wave uncertainty principle applied to matter waves. The classical wave uncertainty principle can be summarized by the intuitively obvious result that as a wave packet is compressed, wavelengths along the compressed direction get shorter.

With respect to aether-drift experiments such as performed by Michelson and Morley, it is clear that if matter waves have the same speed as light waves, then any effect of earth's propagation through the vacuum would equally affect the light waves and the apparatus used to measure them. It has long been recognized that Lorentz covariance of matter is required to explain the null result of such experiments. What has not been generally recognized (though there are numerous exceptions) is that the wave nature of matter provides the basis for relativity and is entirely consistent with the Galilean notion of absolute space and time.

III.1 Waves as particles

Let c represent the characteristic speed of transverse waves in an elastic medium. Recall the Lorentz transformation of wave variables:

$$
\begin{aligned}
\omega' &= \gamma(\omega - \boldsymbol{\beta} \cdot c\mathbf{k}) \\
ck'_{||} &= \gamma\big(ck_{||} - \beta\omega\big) \\
\mathbf{k}'_{\perp} &= \mathbf{k}_{\perp}
\end{aligned}
\tag{30}
$$

Suppose one has a pure oscillation ($\mathbf{k}$=0) in a given frame of reference. Assume as in quantum mechanics that the rest energy ($m_0 c^2$) of a particle with mass m_0 is proportional to angular frequency ω_0:

$$
\hbar\omega_0 = m_0 c^2
\tag{31}
$$

where $\hbar = h/2\pi$ and h is Planck's constant.

In a relatively moving frame in which the velocity moves with speed v, Lorentz transformations would yield this angular frequency, wave number, and wavelength:

$$\omega' = \gamma\omega_0 = \frac{\gamma m_0 c^2}{\hbar} = \frac{E'}{\hbar}$$
$$k'_{||} = \frac{\gamma\beta\omega_0}{c} = \frac{\gamma m_0 v}{\hbar} = \frac{p'}{\hbar} \qquad (32)$$
$$\lambda' = \frac{2\pi}{k'_{||}} = \frac{h}{p'}$$

where $p' = \hbar k'_{||} = \gamma m_0 v$ is the wave momentum along the direction of motion. These equations show the relationships between wave and particle variables.

The wavelength λ' is called the de Broglie wavelength. Its existence leads to interference between massive particles, which is strong evidence that stationary particles consist of standing oscillations. The de Broglie wavelength is observable in interference and diffraction just like wavelengths of ordinary plane waves. For example, electrons passing through a crystal exhibit the same diffraction as x-rays whose wavelength matches the electron's de Broglie wavelength. [Davisson 1927, Thomson 1927] Such interference has been observed not just with elementary particles, but also with large molecules such as buckminsterfullerene C_{60}:

(https://www.univie.ac.at/qfp/research/matterwave/c60/).

Let $M = m_0 c^2/\hbar$. Assume that some wave amplitude $s_i(\mathbf{x},t)$ evolves according to the Klein-Gordon equation:

$$\frac{\partial^2}{\partial t^2} s_i = (c^2\nabla^2 - M^2)s_i \qquad (33)$$

It is common to use Fourier decomposition so that the Klein-Gordon equation can be written as:

$$\omega^2 S_i = (c^2 k^2 + M^2)S_i \qquad (34)$$

where $S_i(\mathbf{k}, \omega)$ is the Fourier transform of the wave amplitude $s_i(\mathbf{x}, t)$. The wave group velocity u, which we take to be the particle velocity, is given by:

$$u = \frac{d\omega}{dk} = \frac{k}{\omega}c^2 = \frac{\left(\omega^2 - M^2\right)^{1/2}}{\omega}c \tag{35}$$

Solving for $k = \omega u/c^2$ using Eqns. 34 and 35:

$$k = \left(1 - \frac{u^2}{c^2}\right)^{-1/2}\frac{M}{c^2}u = \gamma\frac{M}{c^2}u \tag{36}$$

where we have used the familiar definition of γ to obtain the expression on the right.

Substitution into the wave equation yields:

$$\omega = M\left[1 + \gamma^2\frac{u^2}{c^2}\right]^{1/2} = \gamma M \tag{37}$$

Using the definition $M = m_0 c^2/\hbar$, we obtain the quantum mechanical relations for moving particles:

$$\hbar k = \gamma m_0 u \quad \text{(momentum)} \tag{38}$$
$$\hbar\omega = \gamma m_0 c^2 \quad \text{(energy)}$$

$$[\hbar\omega]^2 = \frac{\left[m_0 c^2\right]^2}{1 - u^2/c^2} = \frac{\left[m_0 uc\right]^2 + \left[m_0 c^2\right]^2 - \left[m_0 uc\right]^2}{1 - u^2/c^2}$$
$$= \left[\hbar kc\right]^2 + \left[m_0 c^2\right]^2$$

These equations for wave quantities are consistent with relativistic expressions for moving particles.

III.2 Energy and momentum

Electron waves have energy proportional to frequency ($E = \hbar\omega$) and momentum proportional to the wave vector ($\mathbf{p} = \hbar\mathbf{k}$). Classically, the quantity $\hbar$ must represent the integrated wave amplitude, which is proportional to angular momentum. We assume that all matter waves have similar proportionalities. Using these substitutions in the preceding equations yields the relativistic relations:

$$\mathbf{p} = \gamma m_0 \mathbf{u} \qquad (39)$$
$$E = \gamma m_0 c^2$$
$$E^2 = p^2 c^2 + m_0^2 c^4$$

Hence the quantum mechanical and relativistic expressions for momentum and energy are equivalent.

The last equation, given the first two, merely expresses the tautology:

$$c^2 = u^2 + \left(c^2 - u^2\right) = u^2 + \frac{c^2}{\gamma^2} \qquad (40)$$

However, this equation also has a geometrical interpretation. Consider the velocity triangle in Figure III–1.

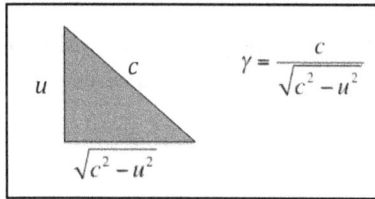

Figure III–1: This is the Pythagorean relation for a right triangle with sides: $(c, u, \sqrt{c^2 - u^2}\,)$.

The hypotenuse c, which corresponds to energy in the last of Eqs. 39, indicates that the disturbance moves with speed c. The velocity v corresponds to momentum and indicates propagation in the direction of the wave vector. The velocity $\sqrt{c^2 - u^2}$ corresponds to mass and indicates propagation perpendicular to the wave vector (or at least independently from the wave vector: the Pythagorean relation also holds, on average, for cycloidal motion u_c

defined by $u_c = \hat{x}_\perp u_\perp \cos\theta + \hat{x}_{||}(u_\perp \sin\theta + u_{||})$ with $u_{||} \to u$ and $|u_c| = c$). Since the propagation associated with mass does not yield any net transport of the disturbance, it must be at least approximately periodic, and the simplest assumption is circular motion. The general propagation of the wave would then be helical or cycloidal (or in between). Hestenes [1990] has similarly interpreted rest mass as indicating circular particle motion at the speed of light.

Multiplying each side of the velocity triangle by $\gamma m_0 c$ yields the energy-momentum relation in Eq. 39. The corresponding triangle is shown in Figure III–2.

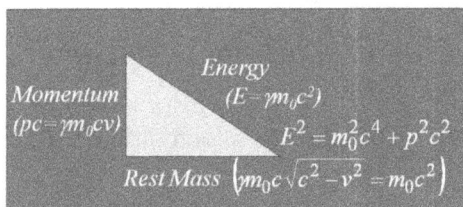

Figure III–2: Triangular relationship between rest mass, momentum, and energy.

If the stationary frequency of an elementary particle is really associated with circular motion at the speed of light, then we can compute the radius of the motion. For electrons we have:

$$R_c = \frac{c}{\omega} = \frac{\hbar c}{m_0 c^2} = 3.8616 \times 10^{-11} \text{ cm} \tag{41}$$

Note that this quantity is different from the Bohr radius ($R_e = \hbar^2/m_e e^2 = 5.2918 \times 10^{-9}$ cm), which is the classical radius of the electron orbit in the ground state of the hydrogen atom. The ratio between these two distances is called the fine structure constant (α):

$$\alpha = \frac{R_c}{R_e} = \frac{e^2}{\hbar c} = \frac{1}{137.04} \tag{42}$$

The velocity and energy triangles provide a simple geometrical interpretation of special relativity, consistent with an interpretation of matter as circulating waves.

III.3   Transformation of velocity

The expression for wave velocity can be combined with the Lorentz transformation laws for frequency and wave vector to work out the transformation properties of the velocity. For relative motion parallel to the velocity only the component $k_{||}$ is affected:

$$u'_{||} = c^2 \frac{k'_{||}}{\omega'} = c^2 \frac{\gamma k_{||} - \beta\gamma\omega/c}{\gamma\omega - \beta\gamma c k_{||}} = c \frac{1 - \beta c/u}{c/u - \beta} \tag{43}$$

$$= \frac{u - \beta c}{1 - \beta u/c} = \frac{u - v}{1 - uv/c^2}$$

This is the transformation law for velocity parallel to the direction of relative motion. For relative motion perpendicular to the velocity the only change is to ω:

$$u'_{\perp} = c^2 \frac{k'_{\perp}}{\omega'} = c^2 \frac{k_{\perp}}{\gamma\omega} = \frac{u_{\perp}}{\gamma} \tag{44}$$

For an arbitrary direction of relative motion $\mathbf{v}$, we use $u_{||} = \mathbf{u} \cdot \mathbf{v}/|\mathbf{v}|$ to obtain the transformation laws for components of velocity parallel $(u_{||})$ and perpendicular $(u_{\perp})$ to the direction of relative motion:

$$u'_{||} = c^2 \frac{\gamma k_{||} - \beta\gamma\omega/c}{\gamma\omega - \boldsymbol{\beta}\cdot\mathbf{k}\gamma c} = \frac{u_{||} - v}{1 - \mathbf{v}\cdot\mathbf{u}/c^2}$$

$$\mathbf{u}'_{\perp} = c^2 \frac{\mathbf{k}_{\perp}}{\gamma\omega - \boldsymbol{\beta}\cdot\mathbf{k}\gamma c} = \frac{\mathbf{u}_{\perp}}{\gamma(1 - \mathbf{v}\cdot\mathbf{u}/c^2)} \tag{45}$$

III.4 Rapidity and Lorentz boosts

Another way to describe particle velocity arises from the observation that in one dimension, solutions to the ordinary wave equation consist of a superposition of forward- and backward-propagating waves:

$$f(x,t) = f_F(x - ct) + f_B(x + ct)$$

We can define a weighted average velocity as:

$$u = \frac{|f_F|c - |f_B|c}{|f_F| + |f_B|} \tag{46}$$

Since the magnitudes are positive-definite, we can express them as exponentials:

$$\begin{aligned} |f_F| &\equiv f_0 \exp(\alpha) \\ |f_B| &\equiv f_0 \exp(-\alpha) \end{aligned} \tag{47}$$

for appropriately chosen α and positive magnitude f_0. The weighted average velocity is then:

$$u = \frac{f_0 \exp(\alpha) - f_0 \exp(-\alpha)}{f_0 \exp(\alpha) + f_0 \exp(-\alpha)} c = c \tanh \alpha \tag{48}$$

The quantity α is called the <u>rapidity</u>.

Starting from rest with $|f_F| = |f_B| = f_0$, a <u>Lorentz boost</u> to velocity $u_1 = c \tanh \alpha_1$ is achieved by multiplying the forward magnitude by $\exp(\alpha_1)$ and multiplying the backward amplitude by $\exp(-\alpha_1)$. Repeating this process for $u_2 = c \tanh \alpha_2$ yields the relativistic velocity addition formula:

$$\begin{aligned} u_{1+2} &= \frac{f_0 \exp(\alpha_1 + \alpha_2) - f_0 \exp(-\alpha_1 - \alpha_2)}{f_0 \exp(\alpha_1 + \alpha_2) + f_0 \exp(-\alpha_1 - \alpha_2)} c \\ &= c \tanh(\alpha_1 + \alpha_2) \end{aligned} \tag{49}$$

Since the hyperbolic tangent varies between $+1$ and -1, the magnitude of velocity can never exceed the characteristic wave speed.

III.5 The twin paradox

One supposedly non-intuitive consequence of relativity is that two twins can change their relative ages through motion. If one twin (Theo=O) remains stationary while the other twin (Primo=O') takes a high-speed journey through space, then the twin who traveled will return younger that the twin who stayed home. This is an example of time dilation. A more common manifestation of this phenomenon is the observation that high-energy cosmic ray particles with relativistic speeds have longer lifetimes than otherwise identical slow-moving particles. Time dilation has even been measured directly (in conjunction with gravitational effects) by flying atomic clocks in airplanes and measuring the time shift relative to a stationary clock. [Halefe and Keating 1972] Although the effect of motion on time may seem almost magical, the explanation is really quite simple.

Consider a clock that counts the number of circular orbits executed by a simplified electron wave. Any clock made of matter waves will tick at a proportionate rate. While the stationary electron executes a circular path, a moving electron executes a helical (or cycloidal) path with the same absolute speed c (see Figure III–3).

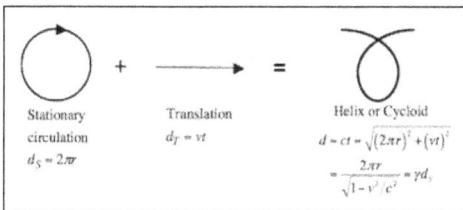

Figure III–3: Time Dilation: Moving matter waves propagate farther than stationary matter waves during each cycle. Therefore, moving clocks tick more slowly than stationary clocks. d_s = distance traveled in one cycle of stationary wave, d_T = translational distance. The distance formula for the cycloid is exact only for an integer number of cycles.

Since the moving electron travels farther than the stationary electron during each rotation cycle, a moving electron clock ($t' = \omega'\tau$) will tick more slowly than a stationary one ($t = \omega\tau$). For a translational velocity of $v_{\parallel}$, the speed of circulation is:

$$v'_{\perp} = \left(c^2 - v_{\parallel}^2\right)^{1/2} = c/\gamma \tag{50}$$

and therefore the moving clock ticks more slowly ($t' < t$) by the factor:

$$\frac{t'}{t} = \frac{v'_{\perp}}{v_{\perp}} = \frac{v'_{\perp}}{c} = 1/\gamma \tag{51}$$

This is equivalent mathematically and similar physically to the derivation above of time dilation for sound waves in water. Hence the moving Primo will age less than the stationary Theo.

We have stated before that wave measurements cannot determine absolute motion relative to the medium. Therefore, Primo should end up younger than Theo even if they are initially moving with respect to the medium. Suppose that the two twins Primo and Theo are initially moving together with velocity v_1 in the x-direction. A stationary observer (O') sees Primo slow to a stop at $t' = 0$, wait for a time $t' = T_1$, then accelerate to speed v'_2 for a time interval T_2 to catch up with Theo at time $t' = T_1 + T_2 = T$. In this case Primo is actually aging more rapidly than Theo at first, but then ages very slowly while trying to catch up. Note that since each travels the same distance in time T, we have $v_1 T = v'_2 T_2$, and therefore:

$$\begin{aligned} T_1 &= T(1 - v_1/v'_2) \\ T_2 &= T(v_1/v'_2) \end{aligned} \tag{52}$$

At the time the twins meet up again, O' sees that Theo has aged by $T'_T = T/\gamma_1$ since his clock is running slower than a stationary clock, using $\gamma_i = \left(1 - v_i^2/c^2\right)^{-1/2}$. But O'' sees that Primo has aged by:

$$T'_P = T_1 + T_2/\gamma_2 = T(1 - v_1/v'_2 + v_1/\gamma_2 v'_2) \qquad (53)$$

Transforming from Theo's reference frame using Eq. 43, the velocities in the relatively moving frame are:

$$v'_1 = \frac{-v_1 + v_1}{1 - v_1^2/c^2} = 0$$

while Primo moves away, and

$$v'_2 = \frac{v_1 + v_1}{1 + v_1^2/c^2} = \frac{2v_1}{1 + v_1^2/c^2}$$

while Primo returns.

Substituting into Eq. 53 for Primo's aging yields $T'_P = T/\gamma_1^2$ in the O' reference frame. Hence O' observes Primo age less than Theo by the factor $1/\gamma_1$, just as was calculated in Theo's reference frame.

Hence the twin who moves away and comes back always ages less than the twin whose motion was constant. This is a direct consequence of the wave nature of matter.

IV. Alternative Interpretations

"It is only the relation of the magnitude to the instrument that we measure, and if this relation is altered, we have no means of knowing whether it is the magnitude or the instrument that has changed." — Henri Poincaré

The reader should be cautioned that the (relatively) simple interpretation of relativity presented here is not generally accepted or even understood. Since its inception at the dawn of the 20th century, the Principle of Relativity has been interpreted as a physical law rather than as a purely mathematical relationship between space and time measurements. It is believed that geometrical relationships between measurements accurately represent the geometry of physical space. Such an interpretation assumes that measurements of distance and time can approach perfection rather than being limited by the nature of matter. The four-dimensional space-time of Special Relativity is referred to as "Minkowski space". According to our point of view, Minkowski space is the set of possible measurements made with waves propagating in a Galilean physical space-time.

It has long been recognized that compliance with the Principle of Relativity requires equations describing matter to be Lorentz covariant. However, the converse logic has been largely ignored. Lorentz covariance is a property of waves in Galilean space-time, and therefore the Lorentz-covariance of equations for matter waves is consistent with a universe evolving in a Galilean space-time. Thus, although absolute motion cannot be measured using light and matter waves, there is no reason to presume that absolute motion has no intrinsic meaning. If another type of wave could be measured (possibly longitudinal waves) then it may be possible to determine absolute motion with respect to an aether.

More realistically, spin angular momentum has a natural interpretation as the intrinsic angular momentum of the aether, whereas orbital angular momentum is associated with wave propagation. [Close 2009] Hence the well-established existence of spin angular momentum could be regarded as confirmation of the existence of an aether. The interpretation of Special Relativity as an intrinsic property of space-time is a philosophical preference that is in no way justified by evidence.

Sir William Thomson (Lord Kelvin) set a high standard for assessing his understanding of physical phenomena: "I am never content until I have constructed a mechanical model of the subject I am studying. If I succeed in making one, I understand; otherwise I do not." Using our conceptual model of matter as waves propagating in circles, we can easily construct a physical model to demonstrate the effect of relative motion.

The circulating wave model of particles on the following page (and on the last page of the text) consists of illustrations of two sets of wave crests as shown in Figure IV-1. The wave crests are drawn as black lines of equal length and are equally spaced along the vertical direction on the page. Each set of wave crests is accompanied by a gray arrow indicating the direction of propagation and the distance traveled in one unit of time. Since the waves are assumed to propagate with equal speed (the speed of light), the gray arrows are of equal length.

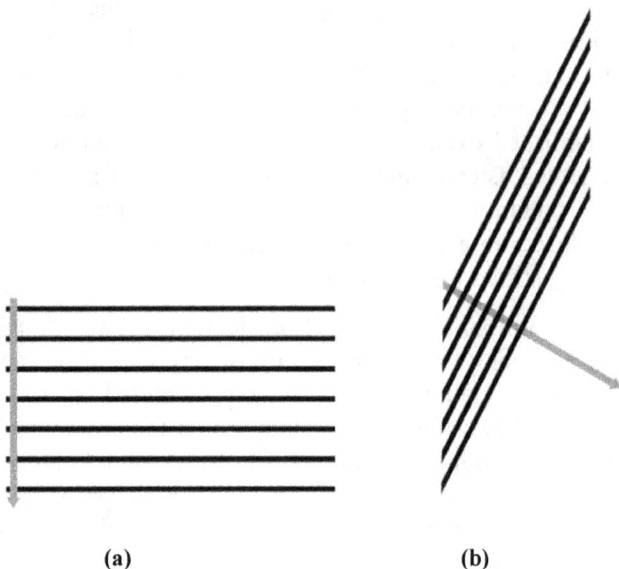

(a)   (b)

Figure IV-1: (a) Model of circular wave propagation with the vertical axis representing the azimuthal direction. (b) Model of helical wave propagation with speed $v = 0.866c$ and $\gamma = 2$. These patterns are designed to be enlarged and copied onto a transparency sheet, then rolled into a cylindrical tube (if printed on ordinary paper, shine a light into the tube to illuminate the wave pattern).

When the sheet is rolled into a cylinder as in Fig. IV-2, the wave packet (a) on the left models a stationary or standing wave (or half of a standing wave) with wave frequency assumed to be $f_0 = m_0 c^2 / \hbar$ and wavelength $\lambda_0 = h/m_0 c$. These relationships are based on the two equations for particle energy: $E = mc^2$ and $E = hf$, where $m = \gamma m_0$ is the relativistic effective mass, m_0 is rest mass, $\gamma = c/\sqrt{c^2 - v^2}$ is the Lorentz factor, c is the speed of light, v is the particle velocity, and h is Planck's constant.

The gray arrow represents the distance light travels in one unit of time, as measured by a stationary observer. The internal clock ticks once each time the wave traverses the circle. If printed such that the length of the arrow is 6 cm, the time for light to make one revolution would be 200 picoseconds.

The model waves on the cylinder look like this:

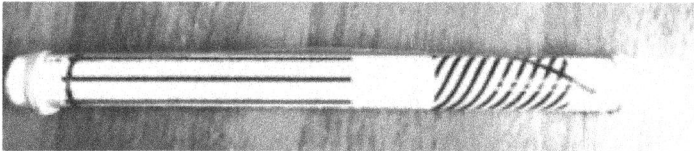

Figure IV-2: Particle wave model rolled into a cylinder.

The theoretical model of particles as circulating waves offers a simple means for understanding special relativity. [Giese 2009, Christie 2016] De Broglie waves satisfying the Klein-Gordon equation in a central potential also propagate in circles, [Synge1954] and the mass term in the Dirac equation can be interpreted as representing circular motion. [Close 2009, Hestenes 1990] This model is a simplification because it restricts the circular motion to a single radius. If the circular motion is assumed to be in another dimension, then this could be a model of string theory.

Rotating the wave crests as in Figure IV-1(b) results in helical wave propagation with average velocity $v = 0.866\ c$ along the axis of the cylinder, and Lorentz factor $\gamma = 2$. The length of the wave crests, the distance traveled in one unit of time (arrow), and the spacing along the circumference of the cylinder are unchanged. The moving wave packet has wavelength $\lambda = \lambda_0/\gamma$ and relativistic frequency $f = \gamma f_0$. This is the usual relativistic frequency shift. Applying the relationship $E = hf$ shows that the moving wave packet has higher energy than the stationary packet. The difference is the relativistic kinetic energy.

The width of the moving wave packet is reduced by a factor of $1/\gamma$. This length contraction was proposed by Fitzgerald and made quantitative by Lorentz in order to explain the null result of the Michelson-Morley experiment. [Fitzgerald 1889, Lorentz 1892, Michelson 1887]

Propagation speed in the azimuthal direction, which measures time, is also reduced by a factor of $1/\gamma$. The arrow showing propagation of the moving packet in one unit of time (as measured by the stationary particle) only travels halfway around the cylinder. This corresponds to half of a tick on the clock, thereby demonstrating relativistic time dilation: a moving clock ticks slower than a stationary clock.

In addition to demonstrating Lorentz transformations, this model also demonstrates an important aspect of quantum mechanics. The distance along the propagation axis between wave crests of the moving wave packet is $\lambda_{||} = (\lambda_0/\gamma)(c/v) = h/(\gamma m_0 v) = h/p$. This is the de Broglie wavelength of a moving "particle". More generally, the de Broglie wavelength results from a Lorentz boost of a stationary oscillation, as discussed previously in Section III.1.

Hence, we have achieved Lord Kelvin's level of understanding for Special Relativity. For a demonstration of this model, see:

http://www.youtube.com/watch?v=tJt6y9ioTg8

With the abandonment of Galilean space-time, physicists dismissed the possibility of a physical, rather than merely mathematical, basis for Special Relativity. They also rejected the modeling methods that successfully yielded Lorentz covariant equations for electricity, magnetism, and light. Yet classical models of disturbances in a solid or mechanical aether have historically produced useful equations consistent with the Principle of Relativity. Perhaps future mechanistic models of the universe will likewise succeed.

References

Bohr, N. (1913) On the Constitution of Atoms and Molecules. *Phil. Mag. S. 6*, **26**(151):1-25.

Boussinesq, J. (1868) Théorie nouvelle des ondes lumineuses, *J. de Mathématiques Pures et Appliquées Sér. II*. **13**:313-339, 425-438.

de Broglie, L. (1924) *Recherches sur la Théorie des Quanta*, PhD Thesis (Paris: University of Sorbonne).

Christie, W. H. F. (2016) Rotating Wave Theory of the Electron as a Basic Form of Matter and Its Explanation of Charge, Relativity, Mass, Gravity, and Quantum Mechanics. url = https://www.billchristiearchitect.com/physics-rotating-wave.

Close, R. A. (2009) A classical Dirac Equation. In *Ether Space-time and Cosmology Vol. 3* (M. C. Duffy and J. Levy, eds., Apeiron, Montreal) 71-83.

Davisson, C. and Germer, L. H. (1927) Diffraction of Electrons by a Crystal of Nickel. *Phys. Rev.* **30**:705–40.

Dyson, F. W., Eddington, A. S., and Davidson, C. (1920) A determination of the deflection of light by the Sun's gravitational field, from observations made at the total eclipse of 29 May 1919. *Philosophical Transactions of the Royal Society* **220**A: 291–333.

Einstein, A. (1905a) Zur Elektrodynamik bewegter Körper. *Annalen der Physik* **17**:891 – 921.

Einstein, A. (1905b) Über einen die Erzeugung und Verwandlung des Lichtes betreffenden heuristishen Gesichtspunkt. *Annalen der Physik* **17**:132-148.

Einstein, A. (1956) *The Meaning of Relativity Fifth Edition* (Princeton: Princeton University Press).

Elsasser, W. (1925) Bemerkungen zu Quantenmechanik freier Elektronen. *Naturwiss.* **13**:711.

Feynman, R. (1969) *The Physics Teacher*, **7** September, 313-320.

Fitzgerald, G. F. (1889) The Ether and the Earth's Atmosphere. *Science* **13**:39.

Giese, A. (2009) Relativity Based on Physical Processes, not on Space-Time. In *Ether Space-time and Cosmology Vol. 3* (M. C. Duffy and J. Levy, eds., Apeiron, Montreal) 143-192.

Hafele, J. C. and Keating, R. E. (1972) Around-the-World Atomic Clocks: Predicted Relativistic Time Gains. *Science* **177**, 166-168.

Heisenberg, W. (1927) Über den anschaulichen Inhalt der quantentheoretischen Kinematik und Mechanik. *Zeitschrift für Physik* **43** (3–4), 172–198.

Hestenes, D. (1990) The Zitterbewegung Interpretation of Quantum Mechanics. *Found. Phys.* **20**(10):1213-1232.

Huygens, C. (1690) *Traité de la lumière.*

Larmor, J. (1900) *Aether and Matter* (Cambridge, University Press).

Lodge, O. (1892) On the present state of knowledge of the connection between ether and matter: an historical summary. *Nature* **46**:164-165.

Lodge, O. (1893) Aberration Problems. *Phil. Trans.* **184**:727-804.

Lorentz, H. A. (1892) La Théorie electromagnétique de Maxwell et son application aux corps mouvants. *Archives Néerlandaises des Sciences Exactes et Naturelles* **25**: 363–552.

Lorentz, H. A. (1904) Electromagnetic Phenomena in a System Moving with Any Velocity Smaller than That of Light. *KNAW, Proceedings, 6, 1903-1904, Amsterdam*, p. 809-831. http://www.dwc.knaw.nl/DL/publications/PU00014148.pdf

MacCullagh, J. (1839) On the dynamical theory of crystalline reflexion and refraction. *Proc. Irish Acad.* **1**:374-9.

Maxwell, J. C. (1861a) On physical lines of force. Part 1. The theory of molecular vortices applied to magnetic phenomena. *Phil. Mag.* **21**:161-175.

Maxwell, J. C. (1861b) On physical lines of force. Part 2. The theory of electrical vortices applied to electric currents. *Phil. Mag.* **21**:281-291, 338-348.

Maxwell, J. C. (1862a) On physical lines of force. Part 3. The theory of electrical vortices applied to statical electricity. *Phil. Mag.* **23**:12-24.

Maxwell, J. C. (1862b) On physical lines of force. Part 4. The theory of electrical vortices applied to the action of magnetism on polarized light. *Phil. Mag.* **23**:85-95.

Michelson, A. A. and Morley, E. W. (1887) On the Relative Motion of the Earth and the Luminiferous Ether. *Am. J. Sci.* (3rd series) **34**:333-345.

Planck, M. (1900) Ueber das Gesetz der Energieverteilung im Normalspectrum. *Verh. deutsch. phys. Ges*, **2**:202-204, 237-245.

Poincaré, H. (1904) L'État actuel et l'Avenir de la Physique mathématique, Conférence lué le 24 septembre 1904 au Congrès d'Art et des Science de Saint-Louis (The present and future of mathematical physics, 24 Sep. 1904 lecture to a congress of arts and science at St. Louis, U.S.A.). *Bull. des Sciences Mathématiques, deuxième Série, tomé* **28**:302-324.

Dr. Suess (1949) *Bartholomew and the Oobleck* (Random House, New York).

Swann, W. F. G. (1941) Relativity, the Fitzgerald-Lorentz Contraction, and Quantum Theory. *Rev. Mod. Phys.* **13**:197-203

Synge, J. L. (1954) *Geometrical Mechanics and De Broglie Waves* (Cambridge University Press, Cambridge) p. 101.

Taylor, B. N. (1995) *Guide for the Use of the International System of Units (SI), NIST Special Publication 811* (Gaithersburg, MD: National Institute of Standards and Technology) Appendix A.

Thomson, G. P. and Reid, A. (1927) Diffraction of cathode rays by a thin film. *Nature* **119**:890-5

Whittaker, E. (1951) *A History of the Theories of Aether and Electricity, vol. 1* (Edinburgh: Thomas Nelson and Sons Ltd.).

Whittaker, E. (1954) *A History of the Theories of Aether and Electricity, vol. 2* (Edinburgh: Thomas Nelson and Sons Ltd.).

Figures

The following figures are believed to be free of copyright restriction and were obtained from the sources listed. Other figures are either original works or are cited in the figure caption.

Figure I–1: Christian Huygens
http://www-history.mcs.st-and.ac.uk/history/PictDisplay/Huygens.html
Figure I–3: Isaac Newton
http://www-history.mcs.st-and.ac.uk/Mathematicians/Newton.html
Figure I–4: Thomas Young
https://mathshistory.st-andrews.ac.uk/Biographies/Young_Thomas/
Figure I–5: Augustin Fresnel
https://mathshistory.st-andrews.ac.uk/Biographies/Fresnel/pictdisplay/
Figure I–6: George Gabriel Stokes
http://www-history.mcs.st-andrews.ac.uk/PictDisplay/Stokes.html
Figure I–7: James MacCullagh
https://mathshistory.st-andrews.ac.uk/Biographies/MacCullagh/pictdisplay/
Figure I–8: Joseph Boussinesq
http://ambafrance-ca.org/HYPERLAB/PEOPLE/bouss.html
Figure I–9: William Thomson (Lord Kelvin, 1824 – 1907)
http://www-history.mcs.st-andrews.ac.uk/history/PictDisplay/Thomson.html
Figure I–9: James Clerk Maxwell
http://www-history.mcs.st-and.ac.uk/history/Mathematicians/Maxwell.html
Figure I–12: Albert Michelson
http://astro-canada.ca/_en/photo690.php?a4313_michelson1
Figure I–14: Hendrik Lorentz
http://www-history.mcs.st-andrews.ac.uk/history/PictDisplay/Lorentz.html
Figure I–15: Jules Henri Poincare
http://www-history.mcs.st-and.ac.uk/history/PictDisplay/Poincare.html
Figure I–16: Albert Einstein
http://www-gap.dcs.st-and.ac.uk/~history/Biographies/Einstein.html
Figure I–17: Max Planck
http://www-gap.dcs.st-and.ac.uk/~history/Biographies/Planck.html
Figure I–18: Neils Bohr
http://www-gap.dcs.st-and.ac.uk/~history/Biographies/Bohr_Niels.html
Figure I–19: Louis Victor de Broglie
http://www-history.mcs.st-and.ac.uk/history/PictDisplay/Broglie.html
Figure I–21: Electron Diffraction
http://en.wikipedia.org/wiki/File:DifraccionElectronesMET.jpg   -   filelinks
(Released under the GNU Free Documentation License GFDL-SELF-WITH-DISCLAIMERS.)

Circulating Wave Model of Special Relativity

Roll sheet around the long axis to see the wave packets. Touch the ends of the gray line on the left (arrow tip to tail). Insert a rolled-up sheet of white paper for better visibility.

Left: Stationary standing wave packet propagating along circular paths ($\gamma = 1$)

Right: Moving wave packet propagating along helical paths ($\gamma = 2$)

Black lines represent wave crests traveling at the speed of light. Both wave packets have the same length of wave crests and the same spacing between crests along the circular direction. The gray arrow represents the distance light travels in one unit of time, as measured by a stationary observer. The internal clock ticks once each time the wave traverses the circle. The moving wave exhibits time dilation (gray arrow only goes halfway around), relativistic frequency increase (wavelength halves), length contraction (wave packet length halves), and the de Broglie wavelength. Let $\theta L_o = m_o c^2$ for the stationary wave. The wavelength along the direction of average motion is then $\lambda_o = \lambda c/v = c^2/(\gamma f_o v) = h/(\gamma m_o v)$

For a more complete explanation, see the book.

The Risky Bices of Special Relativity by Robert A. Close

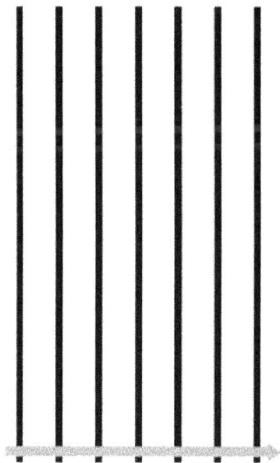

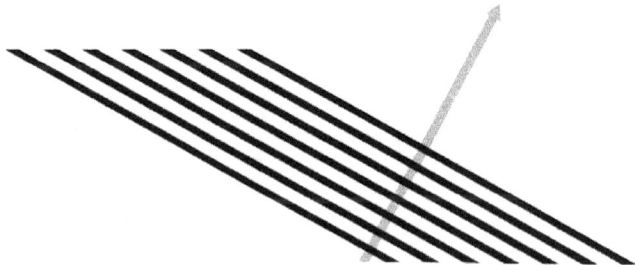